Dedication Page

I dedicate this book to Mr. Mark Dodd, Sea Turtle Program Coordinator for Georgia Department of Natural Resources, and my favorite loggerhead sea turtle hatchling named Chance.

Why I Wrote This Book

I wrote this book to tell kids how they can help animals that are threatened with extinction – animals in this book and other animals I did not mention. If we don't do our part to help the Earth and the animals living here, what kind of world will we leave to future generations?

A

Asian Elephant
Elephas maximas

Animal Facts

Herbivore
Life span = up to 60 years
Group name = herd

Kids like you can help!

Do not buy ivory, so that people will stop
killing them for it. Spread the word!

B

Beluga Whale
Delphinapterus leucas

Animal Facts

Carnivore
Life span = up to 50 years
Group name = pod

Kids like you can help!

Fight global warming. For example, use solar power,
ride your bike, and conserve electricity.

Chimpanzee
Pan troglodytes

Animal Facts

Omnivore
Life span = up to 50 years
Group name = troop

***Kids like you* can help!**

Re-use, reduce, and recycle your cell phones.
Some materials used to make cell phones come
from areas where the Chimpanzee live.

Dugong
Dugong dugon

Animal Facts

Herbivore
Life span = up to 70 years
Group name = herd

Kids like you can help!

Like the Florida Manatee, Dugong eat lots of seagrass, so support efforts to protect these habitats. There may even be an online petition you can sign.

E

Eastern Lowland Gorilla
Gorilla beringei graueri

Animal Facts

Omnivore
Life span = up to 50 years
Group name = band or troop

Kids like you can help!

Give some of your allowance to groups that are teaching local people not to destroy their habitat or kill them for meat.

F

Fin whale
Balaenoptera physalus

Animal Facts

Carnivore
Life span = up to 100 years
Group name = pod

Kids like you can help!

For your next birthday, ask a friend or family member to adopt a fin whale instead of buying you a gift.

Greater
One-horned Rhino
Rhinoceros unicornis

Animal Facts

Herbivore
Life span = up to 40 years
Group name = live alone and not in groups

Kids like you can help!

Teach others about not buying rhino horn products.
Some people think rhino horns can be used as
medicine, but that's not true.

H

Humphead Wrasse
Cheilinus undulatus

Animal Facts

Carnivore
Life span = up to 30 years
Group name = live alone and not in groups

Kids like you can help!

Many fish species are overfished like the Humphead Wrasse, so only eat seafood that can be caught for many years to come. This is called sustainable seafood. Printing a guide can help.

Indus River Dolphin
Platanista minor

Animal Facts

Carnivore
Life span = up to 20 years
Group name = mostly found alone or in pairs

Kids like you can help!

We all need water to live, especially animals like the
Indus River Dolphin. Protect water habitats by doing
things to save water where you live.

Jaguar
Panthera onca

Animal Facts

Carnivore
Life span = up to 20 years
Group name = live alone and not in groups

Kids like you can help!

Jaguar live in tropical jungles, but they also used to be common in the American Southwest. Get involved in efforts to protect jaguar habitat everywhere it can be found.

Little Spotted Kiwi
Apteryx owenii

Animal Facts

Omnivore
Life span = up to 30 years
Group name = mainly live in pairs

Kids like you can help!

Pets such as dogs and cats can kill threatened birds like the Kiwi, so keep them away from bird habitats and teach them not to attack birds.

Ring-tailed Lemur
Lemur catta

Animal Facts

Omnivore
Life span = up to 20 years
Group name = troop

Kids like you can help!

Have a fundraiser to raise money for groups that help manage protected areas of Madagascar where they live.

Manta Ray
Manta birostris

Animal Facts

Omnivore
Life span = up to 20 years
Group name = school

Kids like you can help!

Many ocean animals, including the Manta Ray, can
die from garbage in the water. Never litter!

North Atlantic Right Whale
Eubalaena glacialis

Animal Facts

Carnivore
Life span = up to 100 years
Group name = pod, but can be found alone or in pairs

Kids like you can help!

Many marine mammals like whales are being hurt by chemicals people put in the water. Spread the word, so more people will stop using and dumping harmful chemicals into the ocean.

Bornean Orangutan
Pongo pygmaeus

Animal Facts

Omnivore
Life span = up to 45 years
Group name = live alone and not in groups

Kids like you can help!

Protect their habitat by not using or buying things made with palm oil. Forests are cut down to grow oil palm trees where the Orangutan live.

Tree Pangolin
Phataginus tricuspis

Animal Facts

Carnivore
Life span = up to 20 years
Group name = live alone and not in groups

Kids like you can help!

Do not buy pangolin products or eat them. Instead, buy the book *Roly Poly Pangolin*. Proceeds go to help the Pangolin in Vietnam.

Resplendent Quetzal
Pharomachrus mocinno

Animal Facts

Omnivore
Life span = up to 25 years
Group name = live alone and not in groups

Kids like you can help!

Ask your school science club or class to learn
about the Resplendent Quetzal and adopt
a rainforest to help save their habitat.

R

Red Panda
Ailurus fulgens

Animal Facts

Herbivore
Life span = up to 14 years
Group name = live alone and not in groups

Kids like you can help!

Buy the book *Laloo the Red Panda* or a *Red Panda Endangered Species* chocolate bar. Some of that money is used to help save the Red Panda.

Snow Leopard
Panthera uncia

Animal Facts

Carnivore
Life span = up to 22 years
Group name = live alone and not in groups

Kids like you can help!

Make the Snow Leopard a topic for your school
science project or print a fact sheet to share
with friends and family.

T

Bengal Tiger
Panthera tigris tigris

Animal Facts

Carnivore
Life span = up to 10 years in wild
Group name = live alone and not in groups

Kids like you can help!

Help save the forest habitats of the Bengal
Tiger and other animals by re-using things
made of paper or wood.

Bald-headed Uakari
Cacajao calvus

Animal Facts

Omnivore
Life span = up to 22 years
Group name = troop

Kids like you can help!

Bald-headed Uakari live in the Amazon rainforest, which is being cut down to raise cows. You can help by eating less beef like hamburgers.

V

Vaquita
Phocoena sinus

Animal Facts

Carnivore
Life span = up to 22 years
Group name = pod

Kids like you can help!

The Vaquita is threatened by some fishing practices. Write letters to your local seafood restaurants or stores to tell them why they should only buy and sell fish that are caught in ways that don't hurt the Vaquita.

Whale Shark
Rhincodon typus

Animal Facts

Omnivore
Life span = up to 70 years
Group name = live alone and not in groups

Kids like you can help!

Many people are afraid of sharks, even the Whale Shark. If people learn more about them, they won't be so afraid and will want to protect them. You can help by inviting them to watch a shark science program on TV.

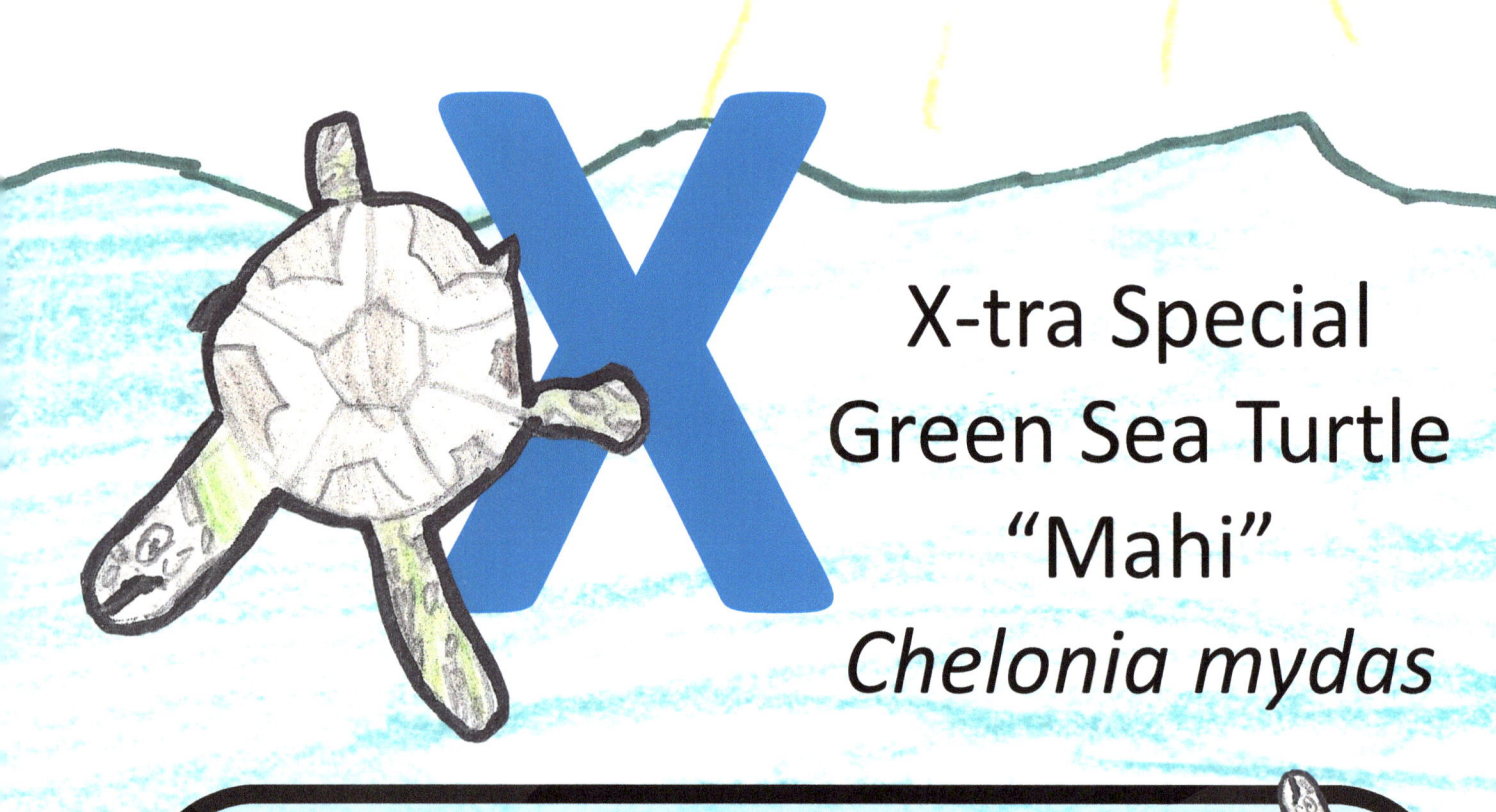

X-tra Special
Green Sea Turtle
"Mahi"
Chelonia mydas

Animal Facts

Herbivore
Life span = up to 80 years
Group name = live alone and not in groups

Kids like you can help!

Mahi lost her front right flipper when it got caught in fishing line. Other animals can be hurt by fishing gear, so never throw or leave fishing gear in the water.

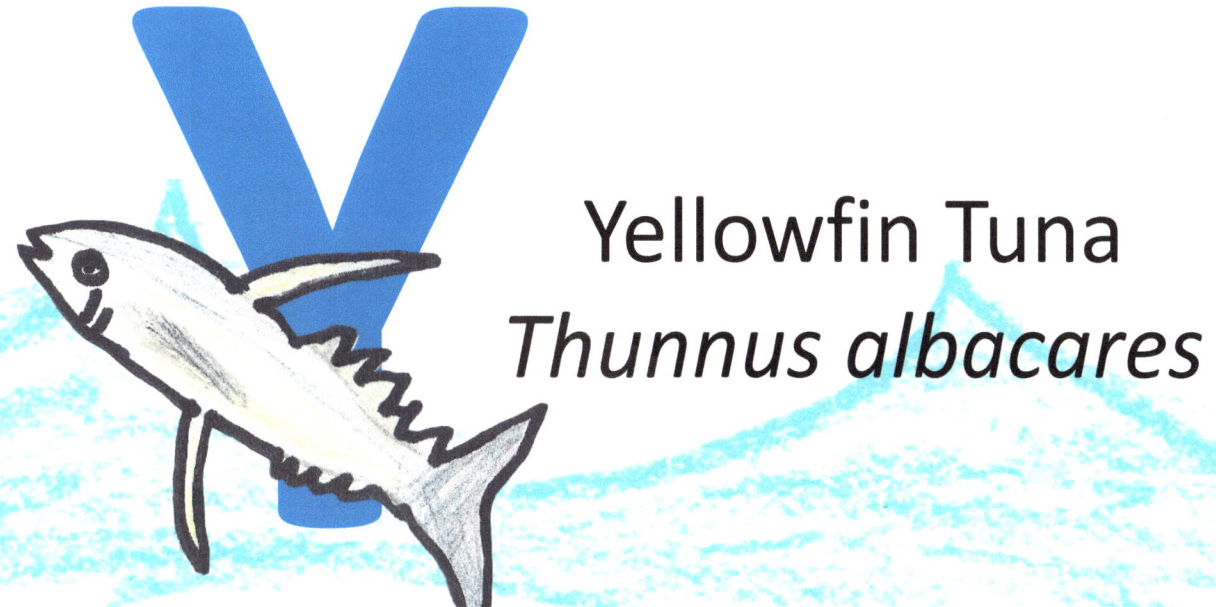

Yellowfin Tuna
Thunnus albacares

Animal Facts

Carnivore
Life span = up to 9 years
Group name = school

Kids like you can help!

The Yellowfin and other tunas are in danger from overfishing, but some tuna companies are trying to change that. Tell your parents to use the canned tuna shopping guide for the best kinds to buy at the store.

Zebra Shark
Stegostoma fascitum

Animal Facts

Carnivore
Life span = up to 30 years
Group name = live alone and not in groups

Kids like you can help!

Do not eat shark fin soup and tell others
why they should not eat it either.

Habitats

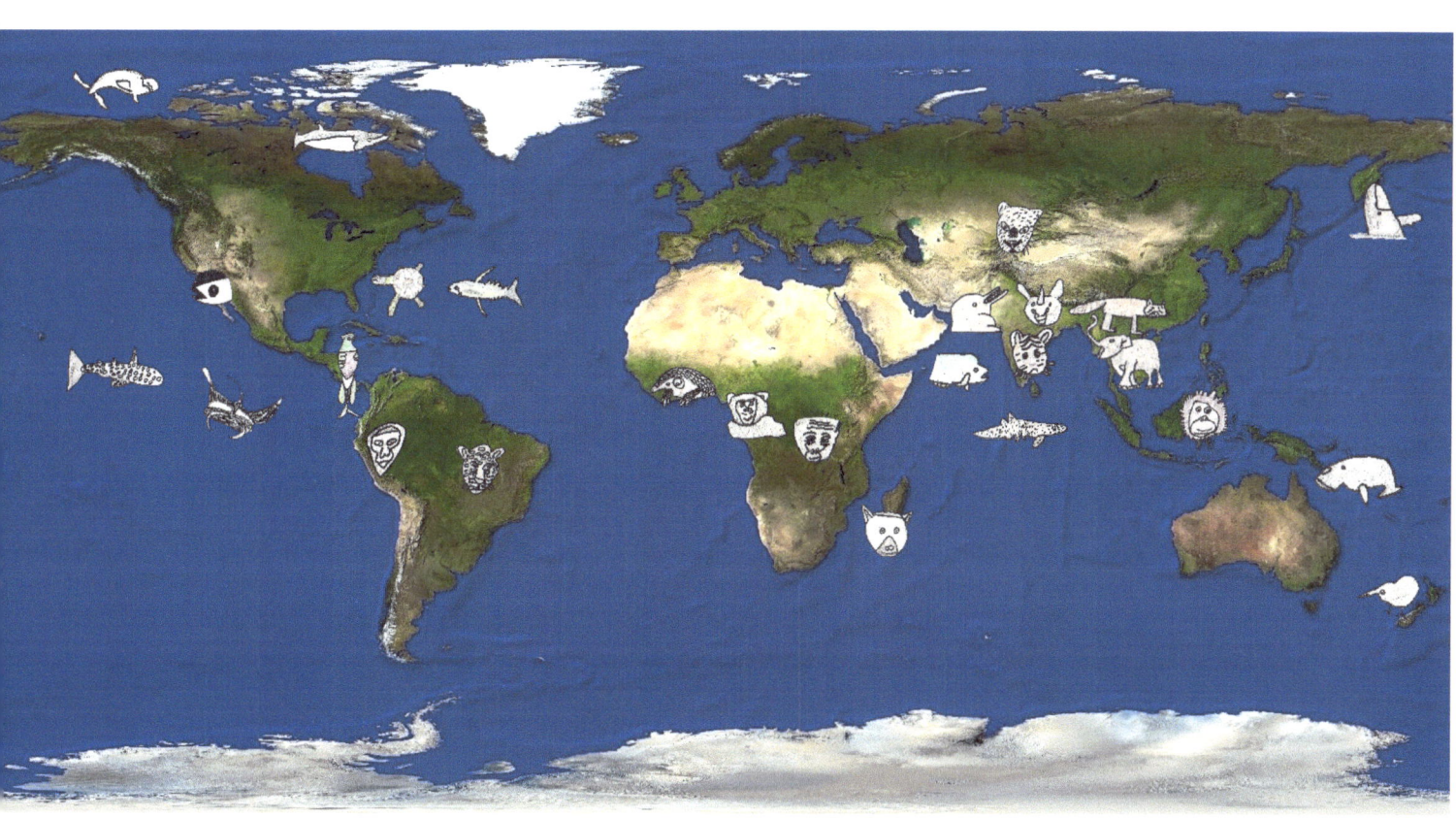

Chance

Chance is my favorite animal. She is a loggerhead sea turtle. Why is she my favorite (because, of course, all sea turtles are cute)? Because she is unique. Let me tell you her story.

Chance

In the summer of 2016, I helped Mr. Mark Dodd, a sea turtle biologist with Georgia Department of Natural Resources, do a sea turtle nest inventory at Sapelo Island, Georgia. When we dug up the nest, we found many of the eggs had not hatched. But some did! In the nest, we also saw two sea turtles that did not make it out of the nest. We kept digging, counting "hatched" and "unhatched" eggs as we went. That's when we spotted a skinny turtle that was still ALIVE! I was able to watch the turtle scamper into the ocean. I named her Chance because now she had a chance to live. In the nest, she wouldn't have survived.

About the Author

This is me!

I am 9 years old and a student at Blackshear Elementary School in Georgia.

I have one younger sister.

I like sharks, sea turtles, and space.

My favorite food is broccoli. I could probably eat broccoli the rest of my life. It's so GOOD!

I like the Florida Gators. F-l-o-r-i-d-a, Go Gators!

Kids like you can
make a difference!

www.ingramcontent.com/pod-product-compliance
Lightning Source LLC
Chambersburg PA
CBHW050411180526
45159CB00005B/2228